WATCHING THE WEATHER

BookLife Rapid Readers

When It's WET

Written by
Noah Leatherland

All rights reserved.
Printed in India.

Written by:
Noah Leatherland

©2024
BookLife Publishing Ltd.
King's Lynn, Norfolk
PE30 4LS, UK

A catalogue record for this book is available from the British Library.

Edited by:
Elise Carraway

ISBN: 978-1-80505-310-1

Designed by:
Jasmine Pointer

All facts, statistics, web addresses and URLs in this book were verified as valid and accurate at time of writing. No responsibility for any changes to external websites or references can be accepted by either the author or publisher.

AN INTRODUCTION TO BOOKLIFE RAPID READERS...

Packed full of gripping topics and twisted tales, BookLife Rapid Readers are perfect for older children looking to propel their reading up to top speed. With three levels based on our planet's fastest animals, children will be able to find the perfect point from which to accelerate their reading journey. From the spooky to the silly, these roaring reads will turn every child at every reading level into a prolific page-turner!

CHEETAH
The fastest animals on land, cheetahs will be taking their first strides as they race to top speed.

MARLIN
The fastest animals under water, marlins will be blasting through their journey.

FALCON
The fastest animals in the air, falcons will be flying at top speed as they tear through the skies.

Photo Credits
Images are courtesy of shutterstock.com. With thanks to Getty Images, Thinkstock Photo and iStockphoto.
Cover — MeSamong, Studio_G, yusufdemirci. Texture throughout — MeSamong. 4–5 — Iakov Kalinin, joerngebhardt68. 6–7 — Martin Ferriz, Pokoman. 8–9 — JoshStocker899, ND700. 10–11 — Tonio_75, DELstudio, Milju varghese, Nadzin. 12–13 — ssuaphotos, Tero Vesalainen, Maike Hildebrandt. 14–15 — Sanja Karin Music, Tom Fern, FANDESIGN. 16–17 — Aluca69, bogdan ionescu. 18–19 — Sisika, Tomsickova Tatyana. 20–21 — CHOKCHAI POOMICHAIYA, Gorloff-KV, Blueastro. 22–23 — nataka, A3pfamily, TORWAISTUDIO.

CONTENTS

PAGE 4	What Is Weather?
PAGE 6	Rain
PAGE 8	When It Rains
PAGE 10	Monsoons
PAGE 12	Helpful Rain
PAGE 14	Floods
PAGE 16	Rainbows
PAGE 18	Fog
PAGE 20	Staying Safe
PAGE 22	Rainy Days
PAGE 24	Glossary and Index

WORDS THAT LOOK LIKE THIS ARE EXPLAINED IN THE GLOSSARY ON PAGE 24.

What Is WEATHER?

The weather is what it feels like outside. It could be hot or cold. It could be wet or dry.

The weather can change throughout the day. It might be warm and sunny in the morning but rainy later on.

RAIN

The air is full of water **VAPOUR**. When water vapour gets cold, it turns into droplets. These droplets make clouds.

RAIN FALLING FROM CLOUDS

The water droplets get bigger and heavier in the clouds. Eventually, they will get too heavy and fall as rain.

7

When It RAINS

Sometimes, rain can fall very lightly. It might feel like there is not much water falling at all.

Other times, lots of rain can fall very heavily. Rain might last for hours or for just a few seconds.

MONSOONS

Some parts of the world get monsoons. Monsoons are strong winds that often bring a lot of rain.

Monsoons can last for months. Monsoons are very important for places that get very hot and dry.

HELPFUL RAIN

Rain is very important to life on Earth. Plants need water to grow. Farmers could not grow food without rain.

Rain also keeps rivers full. People take water from rivers and use it for drinking and washing.

FLOODS

Sometimes, lots of rain creates problems. Too much rain can cause rivers to overflow. This can lead to floods.

In a flood, water spills onto land. Floods can cause lots of damage. Whole roads and houses can get flooded.

RAINBOWS

Rain is needed to create rainbows. Rainbows happen when sunlight hits water droplets in the air in a certain way.

When sunlight hits the water droplets, the light splits apart. Splitting this light makes the colours appear and makes a rainbow.

FOG

Sometimes, wet weather creates fog. Fog looks like a cloud that is very low to the ground.

Fog is made when water vapour collects around dust in the air. Fog usually forms during cold nights.

STAYING SAFE

Wet weather can make things dangerous. Rain can make the ground slippery. Puddles might be deeper than you think.

Heavy rain and fog can make it hard to see. Be extra careful when crossing the road in wet weather.

RAINY DAYS

Wet weather does not mean you have to feel down. Rain is important for keeping the **ENVIRONMENT** healthy.

All you need to do is make sure you wear warm and waterproof clothes. Then, you are good to go!

GLOSSARY

ENVIRONMENT the natural world

VAPOUR tiny amounts of something, such as water, in the form of a gas

INDEX

CLOUDS 6–7, 18
DROPLETS 6–7, 16–17
PLANTS 12
PUDDLES 20
RIVERS 13–14
SUNLIGHT 16–17
VAPOUR 6, 19
WINDS 10